California Natural History Guides: 5

ROCKS AND MINERALS

OF THE

SAN FRANCISCO BAY REGION

BY

OLIVER E. BOWEN, JR.

Drawings by Helen Laudermilk

UNIVERSITY OF CALIFORNIA PRESS
BERKELEY, LOS ANGELES, LONDON

UNIVERSITY OF CALIFORNIA PRESS
BERKELEY AND LOS ANGELES, CALIFORNIA

UNIVERSITY OF CALIFORNIA PRESS, LTD,
LONDON, ENGLAND
© 1962 BY THE REGENTS OF THE UNIVERSITY OF CALIFORNIA
ISBN: 0-520-03244-6 (CLOTHBOUND)
0-520-00158-3 (PAPERBOUND)
LIBRARY OF CONGRESS CATALOG CARD NUMBER: 62-17532
PRINTED IN THE UNITED STATES OF AMERICA

CONTENTS

ILLUSTRATION ON COVER
  Glaucophane-garnet schist (Jenner)

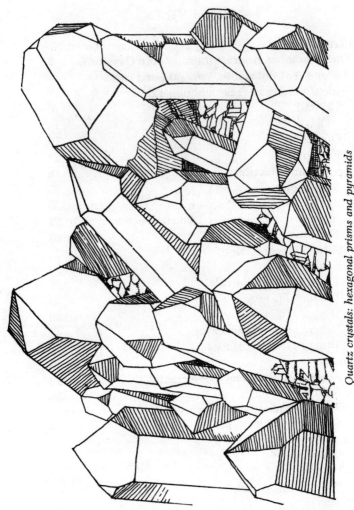

Quartz crystals: hexagonal prisms and pyramids

INTRODUCTION

As the San Francisco Bay region borders the Pacific Ocean and has been inundated repeatedly by the sea, it is not surprising that much of it is underlain by marine sedimentary rocks. Because the local geologic history has been complex, we find also a wide variety of rocks and minerals of other kinds. Indeed, few other heavily populated areas have as great a variety so close at hand. Eruptions of both submarine and terrestrial volcanoes have punctuated the geologic chain of events. Molten rocks of several sorts have been intruded into this segment of the earth's crust. The greatly increased heat and pressure that accompanied these intrusions in some cases completely reformed the adjacent rocks. Hot, mineral-laden waters have risen along fissures from deep within the crust only to stagnate and drop their load of minerals as veins. Along some of our great faults, rocks have been thrust to the surface from deep within the earth's interior. Many sedimentary deposits of sandstone and shale contain abundant remains of former life —the fossil shells of primitive sea animals and the bones of their more complex contemporaries.

In contrast to the relatively stable North Atlantic coastal region, the earth's crust beneath San Francisco Bay is being constantly deformed. Folding and rupture are almost continuous but they work slowly in terms of the human life span. Generally, we are not aware of them until an earthquake takes place, in response to rupture, or a volcano erupts along some fissure. Through these processes of deformation, during which some blocks are elevated and others depressed, and through surface erosion, the variety of rocks present within our crust has been laid bare.

One reason San Francisco Bay region residents have been favored with a look extra deep into the earth's interior is because the San Francisco and Marin peninsulas are transected by a great fracture system known as the San Andreas Fault. This zone of rupturing has been active for at least 20 million years and probably for much longer. Rocks west of the fault have been elevated many thousands of feet and displaced northward many miles as compared to those originally adjacent on the east side of the fault. The continental block west of the San Andreas Fault contains most of the ancient crystalline rocks known in the Bay region—the granites, quartz diorites, schists, marbles, and gneisses.

Folding, faulting, and erosion have also combined to bring to the surface an association of rocks only a little less ancient than the schists and gneisses found west of the San Andreas Fault. These rocks are known as the Franciscan group. They underlie much of Marin and San Francisco peninsulas and form the core of Mt. Diablo. In addition to marine sandstone, shale, limestone, and chert—which make up the bulk of the Franciscan rocks—this group has been invaded by a variety of molten, or at least partly molten, materials that have solidified to form dark igneous rocks. Basalt, diabase, and peridotite, later altered to serpentine, are most common among these. Last and most important, vapors given off during these molten invasions are probably responsible for that highly colorful group of rocks, the glaucophane schists—also part of the Franciscan group.

Would-be collectors of rocks and minerals will do well to note the distribution of the ancient crystalline schists, gneisses, and Franciscan-group rocks, for it is in these that much of the best specimen material occurs.

THE NATURE OF ROCKS, MINERALS, AND CRYSTALS

ELEMENTS, COMPOUNDS, MINERALS, AND ROCKS

Although all things having mass or substance can be converted by natural or man-made processes into energy which is without substance, the materials of our earth can best be considered as consisting of: 1. a single chemical element, 2. one or more chemical compounds, 3. mixtures of several chemical elements. Chemical elements may be defined as the simplest forms of matter that can exist and yet retain, each to itself, a unique set of properties and a definite internal structure. Chemical compounds consist of two or more elements electronically bound together in definite proportions in a particular structure or latticework. Mineral species or varieties are chemical compounds formed by natural processes and consist of one element or one chemical compound. The elemental composition and the structure of this compound may, however, be extremely complex. A rock is simply a more or less cohesive accumulation of one or more minerals, or, in rarer cases—such as volcanic glass—of a mixture of elements having no fixed internal structure.

IDENTIFICATION OF ROCKS AND MINERALS

Minerals may be identified in many ways or combinations thereof among which are general appearance, mineral association, crystal form, chemical composition, cleavability, color, hardness, tenacity, internal structure, behavior in reflected, refracted or transmitted light, and behavior in the presence of an electric current or electromagnetic field. Some minerals, particularly the fine-grained, light-colored varieties, defy identification except by highly-trained personnel using elaborate equipment such as spectrographs, X-ray machines, or high power microscopes. Many, however, are readily identified by their general appearance plus a few simple physical and chemical tests. Identification of others lies between these two extremes. The average layman, after doing some library reading and browsing through labelled collections, can learn to recognize 50 to 100 varieties. Although many hundreds of varieties have been described from various parts of the world, a large proportion of these are rare or are found only in minute amounts in rocks of the San Francisco Bay region.

Minerals commonly are classified and described according to their chemical composition and physical properties. The Dana system of classification, most used throughout the world, is based primarily upon chemical composition, and any student of mineralogy will lean heavily on it for reference. A host of simplified treatises may also be found upon nearly every library shelf. (See the suggested reference list.)

The properties of minerals most useful in the identification of minerals without recourse to specialized equipment are crystalline form, color, streak (color and luster of the powdered mineral), hardness, fracture, luster, tenacity, cleavage, feel, transparency, heft or relative weight, magnetivity, and susceptibility to attack by common acids. The pitfalls inherent in the use of these criteria are explained in nearly all mineralogy books. Inexpensive equipment useful to mineral identification includes a Mohs scale-of-hardness set, penknife, 10-to-20-power hand lens, unglazed porcelain tablet, horseshoe magnet, and eyedropper bottle of dilute hydrochloric, nitric, or sulfuric acid. An acid solution obtained by mixing one part of concentrated acid with three parts of water is satisfactory.

Rocks are conveniently classified into three categories based upon mode of formation—igneous, sedimentary, and metamorphic. Igneous rocks are formed by solidification of molten material. Those that have solidified deep within the earth's crust (intrusive rocks) are apt to consist of intermeshed crystals of minerals readily visible to the naked eye. Those that reach the earth's surface, or close proximity thereto (extrusive rocks), are apt to be finer grained and to consist of visible crystals of various minerals set in a groundmass of natural glass (or of glass and mineral particles too small to be readily visible to the naked eye). Some igneous rocks may consist entirely of glass or glass froth—pumice for example. Igneous rocks are classified according to the proportions of the various constituent minerals and by the textures produced by the disposition of the minerals within the rock. In some cases the chemistry and mode of origin may also enter into the naming of igneous rocks.

Sedimentary rocks are formed through deposition from some transporting medium, such as running water, wind, or gravity. Those transported by water may be further subdivided into fragmental materials that have settled by gravity; materials that have accumulated by chemical precipitation or chemical interaction with bottom materials beneath bodies of water; materials gathered by or under influence of organisms; and materials accumulated by evaporation of mineral-laden water. Sedimentary rocks are classified and identified by their mineral constituents, textures and structures, and their inferred mode of accumulation.

Metamorphic rocks, perhaps the most complex and least understood of the three groups, are formed by reconstitution of other rocks—most commonly under the influence of heat, pressure, and shearing stress deep within the earth's crust. In some cases the original nature of the metamorphic rock can be discerned; in others the original characteristics have been obliterated in process of transformation. Metamorphic rocks can be reconstituted from igneous, sedimentary, or metamorphic parent material, and the transformation almost always is the result of changing environment. Classification of metamorphic rocks is based upon the nature of

[9]

the constituent minerals, the textural and structural characteristics of the rock, and, in some cases, upon the pre-existing nature and manner of transformation of the rock, if discernible.

CRYSTALS

Crystals are the geometrical forms, bounded by plane surfaces, that most minerals assume to some degree during formation from some parent material. The completeness or perfection of crystal formation depends, to a great extent, upon whether space is available for crystal growth and whether there is an adequate source of supply of parent material. Crystal form is determined by the internal structure of the mineral. The internal structure, in turn, is dependent upon the configuration of atoms of the constituent elements and the manner in which they are electrically bonded together. Crystals are divided into six geometrical systems for convenience of classification and description—the isometric, tetragonal, hexagonal, orthorhombic, monoclinic, and triclinic systems. All crystals have fixed mathematical relationships, some of which are quite complex; in fact the science of crystallography is a subject about which volumes have been written. It is dealt with in detail in all mineralogy textbooks, several of which are included in the appended list of references.

MINERALS

QUARTZ FAMILY

The quartz family of minerals are all hard, dense, lightweight materials composed of silicon and oxygen. Quartz in the form of sand, crushed vein quartz or crushed quartzite, is the principal raw material used in the manufacture of glass. Elemental silicon, used in the manufacture of special alloys and abrasives, is obtained from quartz. Colored crystals are cut into gems, and many members of the family are used as ornamental stones in jewelry.

Quartz (silicon dioxide) generally is glassy and virtually colorless, but small amounts of impurities—such as iron, manganese, and titanium—impart color. Quartz crystals, which commonly are well formed and attractive, are hexagonal in cross section, exhibit long prism faces, and are terminated on one or both ends by pyramids. Many, of course, are found only partially formed or distorted, depending on the environment in which they formed. Fragments of quartz crystals, unless faces or parts of faces remain, do not show the planar surfaces (cleavages) exhibited by fragments of feldspar or amphibole. Quartz has a hardness of 7 on Mohs scale, a specific gravity of 2.6 to 2.7, and crystallizes in the hexagonal system. Purple crystals, commonly colored by manganese, are called amethyst. Yellow crystals, colored by iron and possibly several other elements, are called citrine. Rose quartz and smoky quartz are other popular color varieties.

Chalcedony is the microcrystalline form of quartz. It shows no crystal faces or cleavage surfaces, is dense and hornlike on fractured surfaces, and has a hardness about the same as that of a quartz crystal. Specimens that are largely devoid of impurities tend to be light, bluish gray. They have a peculiar greasy feel and cloudy appearance that are distinctive. When impurities are sufficiently great to impart color, the mineral is given a specific variety name. Red or yellow chalcedony, commonly colored by iron oxide, is called jasper; the green variety, colored by nickel or ferrous iron, is called chrysoprase; and the translucent, pale-red variety, colored by ferric iron, is called carnelian. Bloodstone is an impure form, dark green to almost black, containing numerous small patches of red jasper. Moss agate is chalcedony containing black blotches of manganese dioxide.

Some minerals of the quartz family have been given

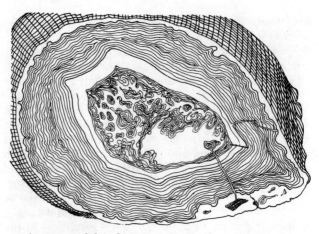

Agate amygdules: thin concentric layers of chalcedony

names based on a distinctive texture or internal structure, such as banding. Most banded varieties have been formed by chemical precipitation from warm mineral waters or water vapors in cavities of a host rock. The host rock is often volcanic. An incompletely filled cavity lined with crystals is called a geode; one which has been completely filled is termed a nodule or amygdule. Agate and onyx are two quartz-family minerals of this type. Agate is made up of concentric bands of chalcedony or alternating bands of chalcedony and opal laid parallel to the walls of the host rock. Onyx also consists of contrasting parallel colorbands of chalcedony, of opal, or of both, but the bands are flat. Agate shells may contain onyx or quartz crystal interiors. Iris agate is a variety laid down in such extremely thin bands that light, passing through thin slices of the mineral, is broken up into the colors of the spectrum.

The third well-known form of quartz-family minerals is opal, a structureless form of silica containing some water. Precious opal, the best-known variety, has a special property resulting from numerous, closely-spaced, shell-like fractures that break up and reflect light in various combinations of irridescent color. The effect produced by opal is similar to that produced by a butterfly's wing or the mother-of-pearl in a seashell. Unfortunately, precious opal has not been found in the San Francisco Bay region. Common opal, which does not reflect light in this fashion, is a frequent associate of chalcedony in bodies of marine chert or in mineral veins. Milk white and shades of brown are the colors most often encountered; less common is the variety hyalite, which is water-clear.

Well-formed quartz crystals are found in white quartz veins cutting light-colored rhyolite on the southeast slopes of Mt. St. Helena, Napa County. The Silverado mine is one of the better-known localities. Other collecting grounds are Fort Point and the Black Hills of San Francisco, where crystals occur in veins in sandstone and chert; the vicinity of the Newman mine on Cedar Mountain, twelve miles southeast of Livermore (yellow crystals); and the basalt flows of Clearlake Highlands. In the latter vicinity glass-clear crystals and crystal groups without crystal faces abound in cavities in the basaltic lava. They are locally known as "Lake County Diamonds."

Nodules and geodes of iris agate and quartz crystals are found in basalt on the ridge above Grizzly Peak Boulevard in Berkeley, between the Fish Ranch Road intersection and the summit of the ridge. Nodules containing agate, onyx, and quartz crystals also occur in masses of light-colored rhyolite along Arlington Avenue in the Berkeley Hills at various places between North Berkeley and the Miramar Golf Course.

Moss agate, chalcedony, and jasper in numerous color shades abound on many pebbly beaches of both San Francisco and Marin peninsulas, notably Pescadero Beach and Half Moon Bay. High-quality beads of carnelian up to the size of a small pea are abundant on the beaches of Fort Cronkhite, Marin County. Streambeds of the main streams draining highlands such as Mt. Diablo, Mt. Tamalpais, and Mt. Hamilton are also good

[13]

Albite feldspar: monoclinic pinacoids and pyramids

collecting grounds. Spherulitic and orbicular jasper
have been found in the country between Sausalito and
the Golden Gate Bridge and at Land's End in San
Francisco. The most famous California localities for
orbicular jasper are Paradise Valley and the Coyote
vicinity in Santa Clara County. There it occurs in veins
and irregular masses in altered serpentine or in stream
beds draining such terrain.

FELDSPARS

Feldspars are among the most common of rock-
forming minerals and sometimes occur in veins and the
vein-like coarse-grained pegmatites. They are silicates
of aluminum and the alkaline earths, specifically
sodium, calcium, and potassium. Feldspars are used
principally in the manufacture of ceramic wares and
glass and are potential sources of aluminum. The idea
that they are the most uninteresting of the minerals is
dispelled when one sees the beautiful white crystal
groups of albite from the Tiburon Peninsula in Marin
County or the handsome green amazonstone crystals
from Pikes Peak in Colorado.

Feldspars are conveniently divided into two groups—the orthoclase or potash group and the plagioclase or soda-lime group. All of the feldspars can be easily cleaved into flat surfaces. Plagioclases are distinguished from the potassium-bearing group by the presence of numerous fine parallel lines or striations on these cleavage surfaces. Plagioclases have been given various variety names based upon chemical composition. Albite has the most sodium; anorthite has the most calcium. Labradorite, intermediate between the two, has nearly equal proportions of sodium and calcium; it is the variety most commonly present in basaltic lava or in the dark granitic rock called gabbro. All feldspars have a hardness of 6 on Mohs scale.

The two familiar potassium-bearing feldspars are orthoclase and microcline. Commonly these are a faint pink or faint salmon color; occasionally they are strongly colored; sometimes they are white or glassy. Microcline is distinguished from orthoclase by the presence of a faint grid of fine wavy bands lighter or darker than the matrix.

Partly-formed crystals of both orthoclase and microclase may be found in coarse-grained pegmatite dikelets cutting the granites of Montara Mountain, San Mateo County, or in like environment in roadcuts along Sir Francis Drake Highway between Inverness and Point Reyes. Many Bay region residents will remember seeing large, pinkish-white crystals in the granites of the Point Lobos area in Monterey County; these are microcline.

By far the finest feldspar specimens obtainable in the Bay region are the white clusters of tabular albite crystals found with lawsonite in veins cutting sandstone at Reed's Station on the Tiburon Peninsula in Marin County. Many are half an inch long. The associated lawsonite shows no striations and tends to be blotched with bluish gray. Otherwise it resembles feldspar. The coarse-grained pillow basalt found in roadcuts along

Sir Francis Drake Highway just west of San Geronimo contains glassy labradorite crystals up to an inch long.

MICAS AND CHLORITES

These minerals are characteristically flaky and soft enough to be scratched by the fingernail. Micas are common constituents of both igneous and metamorphic rocks and survive the abrasive processes of erosion well enough to be found also in detrital sediments. Transparent micas have long been used for insulating electrical equipment and in optical instruments.

Complete crystals of both micas and chlorites are slightly barrel-shaped and are pseudo-hexagonal in cross section even though they belong in the monoclinic system. Whole crystals, however, are difficult to separate from the mesh of matrix crystals in which they occur, so that both micas and chlorites are generally seen as gleaming cleavage surfaces—the "fool's gold" of placer miners. In addition to being soft, the flakes are flexible and transparent or translucent.

The micas are silicates of aluminum together with such other metallic elements as potassium, sodium, magnesium, and iron. The atom groups are arranged in layers, and cleavage is easy because of the layered internal structure. The three best-known mica varieties are muscovite, the potassic, colorless, transparent form; biotite, which is black; and phlogopite, the bronzy magnesium mica most commonly found in marble close to granite intrusions. A fourth variety, mariposite, is familiar to Californians because it is conspicuous in gold ores of the Sierran gold belt and in some altered serpentines of the Coast Ranges. The latter variety is apple green or chrome green, nickel or chromium providing the color. Mariposite is an alteration product of serpentine rock.

The chlorites are similar to micas, but differ slightly in both physical and chemical properties. Most varieties are green or blackish green but of a different shade

[16]

than mariposite. The varieties are indistinguishable except by rather elaborate tests. Chlorites are most commonly seen in metamorphic rocks where they are alteration products of biotite, amphiboles, and pyroxenes.

Micas and chlorites are widely distributed throughout the San Francisco Bay region, but in most places are disseminated as small flakes among other rock-forming minerals. Biotite and muscovite may be found in large flakes or small sheets in pegmatite dikes cutting the Montara Mountain granites of San Mateo County. Crystals and cleavage flakes of muscovite, biotite, and chlorite of thumbnail size are widely distributed through the glaucophane schists of the Reed's Station vicinity on Tiburon Peninsula, Marin County, as well as in similar rocks of the Healdsburg vicinity of Sonoma County. Mariposite is occasionally seen in the "quicksilver rock" of the New Almaden vicinity, Santa Clara County, and in the vicinity of the Mt. Diablo quicksilver mine of Contra Costa County.

AMPHIBOLES

The amphiboles are a lustrous, colorful group of minerals occurring in a wide variety of rocks. Most commonly they are found in long, slim, well-formed crystals having a diamond-shaped cross section. All of the varieties have a hardness of 6 on the Mohs scale and are somewhat heavier than feldspars or quartz-family minerals. With the exception of tremolite, which is white to nearly colorless, the varieties are strongly colored in shades of green, blue, brown, or black. All varieties split into flat cleavage planes of several orientations; two of these commonly meet at angles of either 56 or 124 degrees. The amphiboles have a complex internal structure and chemical composition, being silicates of many metals, among which calcium, iron, magnesium, sodium, and aluminum are, perhaps, most common. They all form in the presence of water vapor, and most varieties contain combined water.

[17]

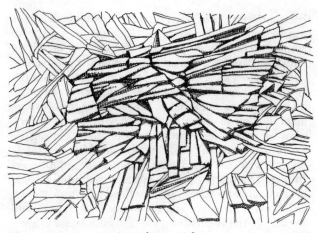

Actinolite crystals

Five varieties, all found in the San Francisco Bay region, are worthy of mention. *Hornblende,* the black, iron-rich variety, is the best-known and is found in the widest variety of rocks. It is an abundant constituent of many granites and volcanic rocks, as well as of schists and gneisses. In the San Francisco Bay region, good large-crystal specimens of hornblende are easily obtained from the glaucophane schists of Tiburon Peninsula, Healdsburg, Mt. Diablo, the Arroyo Mocho of southeastern Alameda County, and near the head of Cutting Boulevard in Richmond. Red garnet, dark blue glaucophane, green actinolite, flaky-white talc, wedge-shaped crystals of honey or pink sphene, and flaky chlorites and micas are common associated minerals.

Actinolite is a deep-green, bladed variety of amphibole, colored by iron in the ferrous condition. It is one of the most conspicuous minerals in the glaucophane schists where it occurs in large masses of crisscrossing, two- or three-inch long crystals associated with pearly talc. Two of the better localities are the Berkeley Hills below Arlington Avenue, from El Cerrito north to Macdonald Avenue in Richmond, and the Tiburon Peninsula at and north of Reed's Station. Most of the localities mentioned for hornblende yield actinolite and glaucophane as well.

Glaucophane, the mineral for which a whole suite of rocks has been named, is a deep-blue, sodium-and-iron-bearing amphibole, common in California and New Zealand but rare elsewhere in the world. It forms as a metamorphic product of many other rocks, such as sandstone, shale, chert, and basalt. Consequently, it accompanies a wide variety of minerals—more than twenty having been identified in a single hand specimen. The most striking specimens of this mineral—bundles of silky-blue, needlelike crystals associated with jadeite, lawsonite, and clinozoisite—to be found in the Bay region are near Valley Ford in Sonoma County. Tiburon Peninsula, Healdsburg, River's End (mouth of the Russian River), and Cazadero are a few of the many other fine localities for glaucophane.

Riebeckite is another blue amphibole slightly different from glaucophane in color, chemical composition, and atomic structure. Glaucophane is deep blue with a violet cast, whereas riebeckite is Prussian blue or blue black. Both contain iron and sodium, but in different proportions. Riebeckite is found in crisscrossing needles or radial acicular groups in the rhyolites of the Valley of the Moon country near Glen Ellen in Sonoma County. The Valley of the Moon flagstone quarry is the best-known locality, but the ridge east of Highway 12 from Glen Ellen to Kenwood affords many fine sites. Riebeckite also occurs in veins cutting glaucophane schists, but there are no notable occurrences in the Bay region.

Tremolite is the iron-free, calcium-magnesium variety of amphibole found most commonly in altered magnesian limestone near granitic intructions. It is off-white to nearly colorless and generally occurs in bundles and radial groups of needlelike crystals associated with calcite. Other likely mineral associates are actinolite, garnet, diopside, idocrase, and phlogopite. The most accessible collecting localities are the hills of Richmond and El Cerrito where tremolite occurs in numerous places with green actinolite. It is also found sporadically in the Santa Cruz marble quarries.

The various members of the pyroxene group are analogous to the amphiboles, but they form at a higher temperature and in the absence of water vapor. Crystals are apt to be short and stubby. Some varieties crystallize in the orthorhombic system, others in the monoclinic system. They are easily cleaved, angles between cleavages being 87 or 93 degrees as compared to 56 and 124 degrees for amphiboles. Pyroxenes are found in many dark-colored igneous rocks, particularly gabbro, peridotite, and basalt. Several varieties are found almost exclusively in metamorphic rocks.

Augite, the black or dark-brown variety containing iron, calcium, and magnesium, is widely distributed through the San Francisco Bay region in such rocks as basalt, andesite, and gabbro. Nowhere in this vicinity, however, does it occur in large or well-formed crystals that are worth collecting.

Diopside, the calcium, magnesium-bearing variety, is usually green to blue green. Although diopside is widely distributed in rocks such as andesite, diorite, and some basalts and gabbros, the best crystals generally are found with calcite in crystalline limestone situated near a granite intrusion. The best collecting localities are the marble quarries in and near Santa Cruz, but fair specimens are sometimes obtained from the glaucophane schists of the Richmond Hills. A rare lilac-colored variety has been found at Fort Point, San Francisco, in veins cutting a green metamorphic rock. Both massive jadelike veins and well-formed crystal groups have been described from this locality.

Jadeite, the sodium-aluminum variety best known in the form of semiprecious jade, is generally some shade of green. It is the hardest (H=6-7) of the pyroxenes and the most sought after by lapidarists. The best locality for coarse-crystalline jadeite is Valley Ford, Sonoma County, where it occurs with glaucophane. It is also found with nephrite jade and associated metamorphosed sandstones at Massa Hill, Marin County, although the specimen material is not of the best quality. Jadeite has also been found along the Russian River near Cloverdale.

Enstatite is the common, green, high-magnesium pyroxene seen in peridotite. As most peridotite masses in the San Francisco Bay region have been altered to various serpentine minerals, fresh enstatite rock is uncommon. However, in many pieces of serpentine rock, the form of the enstatite crystals has been retained throughout alteration and the shiny cleavages are faithfully preserved in the mineral bastite (a form of serpentine). It is common in serpentine of both San Francisco and the Berkeley Hills.

Aegirite is a relatively rare green-to-black pyroxene occurring with blue-black riebeckite at the Valley of the Moon flagstone quarries near Glen Ellen, Sonoma County. It is essentially a sodium-iron silicate. It occurs in stubby crystals or crystal groups in cavities in rhyolite.

Omphacite is a grass-green pyroxene found associated with red garnets at Reed's Station, Tiburon Peninsula. This rock is called eclogite. Omphacite is also found at River's End, Sonoma County, northwest of Jenner.

Diallage is a relatively soft pyroxene (hardness of 4 on Mohs scale) characterized by a lamellar structure and strong tendency to split along the lamellae (thin bands). The cleavage planes commonly have a bronzy cast. Crystals of diallage up to several inches long occur in a pyroxenite pegmatite on the north side of Fort Ross—Monte Rio road, Sonoma County. A coarse-grained diallage-bearing gabbro is found along Bagley Creek one and one-half miles north of Mt. Diablo, Contra Costa County.

[21]

GARNETS

Garnets are a colorful group of very hard, heavy minerals often cut into semiprecious gems or used crushed as an abrasive. They crystallize in the isometric system, range in hardness from 6½ to 7½ on Mohs scale, and show no cleavage. Red, green, and lilac colors probably predominate, but pink, yellow, and orange hues are not uncommon. Garnets are seen most frequently in metamorphic rocks, particularly micaceous schists and crystalline limestone, but they occasionally form in granite or granite pegmatite or even in cavities in rhyolite lava. Well-formed crystals are the rule rather than the exception. Twelve-sided (dodecahedral) or 24-sided (trapezohedral) crystals are most common, but many show intricate combinations of faces. Garnets range widely in composition, being silicates of two or more of the following metals: aluminum, iron, calcium, magnesium, manganese, chromium, and nickel.

In the San Francisco Bay region garnets are found most abundantly in eclogite or glaucophane schist, these two rocks being closely associated at several localities. The Tiburon Peninsula, Healdsburg, mouth of the Russian River, and the Arroyo Mocho of Alameda County are good collecting grounds for red garnets.

Garnet: dodecahedral crystals

Common Olivine is found abundantly in basaltic lavas and is the principal parent mineral from which serpentine forms. The yellowish-green variety, *forsterite*, is occasionally found in small grains in the crystalline limestones of Santa Cruz and Monterey counties. Small crystals of green olivine may be seen in many specimens of basalt picked up in the Berkeley Hills or taken from the dark lava flows of Napa and Sonoma counties. Really good collecting localities have not been described in the San Francisco Bay region although good material has been reported to occur on a ridge west of Coyote Dam, Santa Clara County.

Common olivine, also called chrysolite or peridot, is nearly always green or yellowish-green. It is a silicate of magnesium and iron, iron in the ferrous state causing the color. Olivines have no cleavage, which distinguishes them readily from pyroxenes, and are nearly as hard as quartz (6½ to 7 on Mohs scale). Well-formed crystals commonly have short, flat side-faces mounted above and below by pyramids. The crystallization is orthorhombic. More often than not olivine crystals are not bounded by crystal faces but appear as oval grains. Olivines alter rather easily to green serpentine, reddish iddingsite, or yellowish bowlingite. Presence of these strongly-colored alteration products is often an aid in identification of olivine.

Forsterite, the magnesium-rich variety with little or no iron, is found principally with dolomite and calcite in magnesian limestones situated close to a granite intrusion. It is generally pale yellow or yellowish green. Forsterite is distinguished from green spinel by being much softer and from green diopside by having no cleavage.

Fayalite, the iron-rich, dark-brown-to-black variety, is occasionally found in crystals lining gas cavities in lava flows. It is not known to occur in the Bay region.

Unflawed olivine crystals of good size and color are

[23]

sold as gem stones under the name peridot. Most gem peridots are common olivine found as linings in the gas holes of lavas. One of the better-known localities is the vicinity of Roosevelt Dam on the Salt River in Arizona. Olivine is also used as a refractory material for furnace linings, but for this purpose it is generally manufactured from serpentine.

SERPENTINES AND TALCS

The serpentine and talc groups of minerals are hydrous magnesium silicates that nearly always form as alteration products of pre-existing minerals. The parent materials most commonly are olivines, pyroxenes, and silicious dolomite rock. The Coast Ranges, Sierra Nevada, and Klamath Mountains all contain huge masses of serpentine rock derived by hydration of the parent peridotite, a rock originally made up predominantly of common olivine and various pyroxenes. Many masses of serpentine rocks are found within the San Francisco Bay region and are among our most interesting and economically important rocks. Both chrysotile asbestos and chromite occur in serpentine rock, and many serpentine masses are potential sources of nickel as well. Semiprecious serpentine is used in costume jewelry and at various times has been sawed and polished for building facings. The talc group, which commonly forms as alteration products of serpentine by loss of water and addition of silica, are used in various fired mixtures for wall tile and special heat-resistant ceramicware and in finely pulverized form in cosmetics, pharmaceuticals, mineral fillers, and insecticides.

The two most common and best-known serpentine minerals are *antigorite* and *chrysotile asbestos*. Serpentine rock is made up chiefly of these minerals. *Antigorite* or common serpentine ranges in hardness from 3 to 4, and in specific gravity from 2.3 to 2.6. It is granular or platy and seldom forms good crystals. Specimens of common serpentine found in the San Francisco Bay

[24]

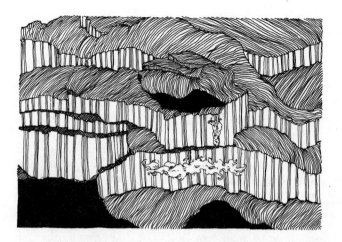

Chrysotile asbestos veinlets in serpentine rocks

region generally show numerous, polished, curved surfaces that have a slick, greasy feel and a high luster. Bastite is the platy subspecies or subvariety of antigorite that takes the form of the enstatite or diallage crystals from which it formed. Picrolite is a hard, brittle-fibrous form prized for mineral specimens.

Chrysotile, the most widely known and used variety of asbestos is made up of innumerable bundles of hair-fine, astonishingly strong, very flexible fibers. It occurs in veinlets cutting a matrix of other serpentine minerals. When the fibers grow perpendicular to the vein walls, the asbestos is called cross-fiber; when they grow parallel to the walls, it is termed slip-fiber. Because of its high melting point, insulating quality, and fiber strength, asbestos is used in fireproof fabrics, heat insulation, and electrical insulation materials.

The *talcs* are massive-to-platy, very soft, sectile, easily pulverized minerals with a greasy or soapy feel. Plates of the mineral are flexible, but not elastic. *Steatite* talc is the very pure, massive, compact variety used in high-grade cosmetics and ceramicware. Most commonly it is white or pale green and carves easily with any edged instrument. *Soapstone* is a massive or strongly foliated impure variety that may be almost any color, depending upon what impurities are present. Iron is the most common of these. Massive soapstone is prized for sculpturing and for building blocks and pulverized soapstone is used as a lubricant, filler, and insecticide carrier. *Crystalline talc* comes in lustrous, pearly, pseudohexagonal plates that really belong to the monoclinic crystal system.

Almost any of the varieties of serpentine mentioned here can be found somewhere in any large serpentine mass. Red Mountain, Alameda County, the vicinity of Joaquin Miller Park in Oakland, the hills back of Richmond and El Cerrito, the east slopes of Mt. Diablo, and the vicinity of the new Mint building in San Francisco are all good places to look. There are many others in almost every county of the San Francisco Bay region.

Crystals and flakes of talc of fair size occur associated with a mesh of dark-green actinolite crystals in fields near the head of Cutting Boulevard, Richmond. The Tiburon Peninsula is also a good area for actinolite-talc rocks. Small masses of light-colored talc are commonly found in most of the serpentine masses throughout the Bay region.

Epidotes

Two members of the epiodote group of minerals are widely distributed through the San Francisco Bay region. *Common epidote* is a brittle, easily-cleaved, deep pistachio-green mineral generally found in well-formed groups of slender crystals. The coloring matter is iron in the ferrous state. *Clinozoisite*, which contains little or no iron, is gray or pale green and does not crystallize quite as vigorously as its sister variety. The epidotes are found in a wide variety of metamorphic rocks and in mineral veins. The best specimens come from crystalline limestones close to a granite intrusion or from veinlets cutting the glaucophane schists. Like the amphiboles, epidotes form only in the presence of water vapor, and water enters their chemical composition. They are silicates of calcium, iron, and aluminum, each of the several varieties predominating in one or another of these metals. Small amounts of manganese, magnesium, or chromium are present in some varieties. Although the epidotes make handsome specimens, no commercial use thus far has been made of any of them.

Both common epidote and clinozoisite are found in crystalline limestones where they are commonly associated with calcite, quartz, garnet, and diopside. In the glaucophane schists glaucophane, albite, micas, actinolite, and quartz are more likely to be associated. The epidotes range in hardness from 6 to 7 on Mohs scale and in specific gravity from 3.2 to 3.5. They crystallize in the monoclinic system.

Piedmontite, the rose-red, manganiferous variety, and allanite, the dark-brown, radioactive variety containing small amounts of cerium, thorium, and yttrium, are found disseminated in small grains in the granites and some of the metamorphic rocks of the Bay region, but good specimens are rare.

THE CARBONATE MINERALS

Calcite, magnesite, rhodochrosite, siderite, dolomite, and ankerite are the principal members of the carbonate group found locally. All of these crystallize in the hexagonal system, cleave readily into rhombic fragments, are easily scratched by a knife blade, and are decomposed by acid with effervescent evolution of carbon dioxide gas. They are composed of one or two metallic elements combined with carbon and oxygen. Because these minerals are rather easily broken down into other chemical compounds, nearly all of them are useful to man. *Calcite*, calcium carbonate, the principal mineral of limestones, is the most widely used of the chemical raw materials. *Dolomite*, the double carbonate of calcium and magnesium, is the principal Cal-

Calcite crystals: dogtooth spar

ifornia source of magnesium chemicals and refractories. Magnesite has been used similarly, and rhodochrosite and siderite, when found in quantity, are ores of manganese and iron, respectively. Nearly all of the carbonate minerals have been used upon occasion as sources of carbon dioxide gas. *Calcite,* the most familiar of the carbonate minerals, is white, colorless, or light-colored unless badly contaminated by impurities. It is widely distributed in mineral veins, in cavities in all sorts of rocks, in sedimentary and metamorphic limestones, and as the cementing agent in many sandstones and pebble conglomerates. Calcite probably has the widest variety of crystal shapes of any mineral, the best specimens coming from veins and cavities where there has been ample space for growth of crystals. The best-known calcite crystal forms probably are the rhomb and the long-pointed, tooth-like variation called dogtooth spar. Crystals of the water-clear transparent variety, Iceland spar, are sought for use in optical instruments.

Calcite-collecting localities are numerous in every San Francisco Bay county. Veins cutting the ancient greenstone basalts of Mt. Diablo's North Peak and the greenstones seen along Highways 1 and 101 north of Golden Gate Bridge; the glaucophane schists of Tibu-

[28]

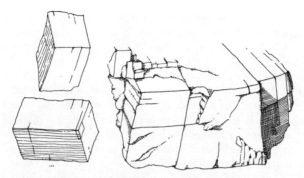

Calcite cleavage rhombs

ron Peninsula, Healdsburg, Richmond, and the Arroyo Mocho; veins in the Fort Point vicinity of San Francisco; and amygdules in the basalts of the Berkeley-Oakland Hills are some suggested places to look.

Dolomite, the second most widely distributed of the carbonate minerals in this vicinity, looks much like calcite but reacts less vigorously to dilute acid than calcite, and well-formed crystals characteristically show curved crystal faces. It occurs in the same kinds of rocks and in the same environment but is much less common than calcite. Veins cutting the rust-colored silica-carbonate rocks that occur as alteration products of serpentine yield the better dolomite specimens found locally. The vicinities of the New Almaden and Guadalupe quick-silver mines are good hunting localities.

Magnesite, the white magnesium carbonate, is found in the San Francisco Bay region exclusively in veins or cauliflower-shaped nodules in serpentine rock. The masses of magnesite are dense and porcelaneous, and any cleavages and crystal faces that might be present are too small to be seen without the aid of a powerful microscope. Well-crystallized specimens are found in the Hillsdale mine south of San Jose. The most notable occurrences of magnesite in porcelaneous form are in the vicinity of the Bald Eagle mine on Red Mountain astride the Stanislaus-Alameda County line east of Mt. Hamilton. The mineral was mined there during World

War II for use in quick-setting magnesia cements. Other magnesite localities are the serpentine masses near Cloverdale, Cazadero, Guerneville, and Cedar Mountain.

Rhodochrosite, the rose-red carbonate of manganese, is by far the most handsome of the carbonate minerals. It is found in several Bay region manganese mines where it is commonly associated with black oxides of manganese, resin-yellow bementite (a manganese silicate), and pink rhodonite (also a silicate of manganese). The occurrence of these minerals is almost entirely confined to thin-bedded cherts (a fine-grained chalcedonic rock) of the Franciscan formation. The most notable local collecting area is the Ladd-Buckeye mining district astride the Alameda-Stanislaus County line in Corral Hollow, roughly midway between Livermore and Patterson. Rhodochrosite is also found in prospect holes on the Miller Ranch at the extreme southeast point of the Los Buellis Hills, Santa Clara County; in manganese prospects six miles west of Cazadero, Sonoma County; and in a series of pits seen along the Arroyo Mocho road east of Mt. Hamilton.

Ankerite, a carbonate of calcium, magnesium, and iron, is a gray-to-tan variety seldom found in good crystals and is commonly overlooked. A common Coast Range environment is rust-colored silica-carbonate rock or "quicksilver rock," as it is known among the miners. Dolomite, quartz, and other carbonate minerals are common associates. Because of its iron content, ankerite often weathers a pale rust-color, and this is its most distinguishing feature. Identification of some fresh specimens can only be determined by chemical tests.

THE METALLIC ORES

Few people think of the San Francisco Bay region as a mining district, yet some of our mines have been among the most lucrative in California. The yield of the New Almaden quicksilver (mercury) mine near

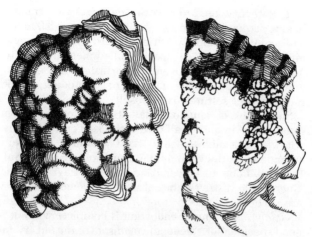

Strawberry and stalactitic forms of cinnabar

San Jose amounts to over $50,00,000 and the monetary value of production of six or eight more mines runs into eight figures. Although the leading metal is quicksilver, there have been times when Bay region chrome and manganese deposits were important to the nation's economy. One mine, the Palisade, near Calistoga, produced substantial quantities of gold, silver, copper, and lead. Small amounts of gold and platinum have been recovered from beach placers on San Francisco Peninsula.

Cinnabar, the bright-red sulfide of mercury is among the most attractive of the ore minerals. Long before the advent of Caucasians to California, cinnabar had been mined and used for ornamental body paint by the Indians, its toxic characteristics notwithstanding! If cinnabar were harder, it would make admirable gem stones as it has an extremely fine color and luster, the ability to refract light strongly, and that aura of romance that subtly enhances value. Although it is generally found in the massive state, cinnabar does form good crystals belonging to the hexagonal system. Rhombic and pyramidal crystals are perhaps most common. Although

[31]

soft, 2 to 2½ on Mohs scale, cinnabar is extremely heavy, having a specific gravity ranging from 8.0 to 8.2. The color occasionally varies toward gray or brown, and there is a black subspecies called metacinnabarite. Mercury is easily recovered from its ore by roasting and distillation of the gaseous roast product; liquid mercury and sulfur dioxide result.

Specimens of cinnabar may still be obtained from mine refuse in almost all of the quicksilver mining districts. Active mines such as the Guadalupe near New Almaden, the Buckman mines near Cloverdale, the Sulfur Bank mine on Lower Lake, and the Mayacmas and Guerneville district mines of northern Sonoma County are particularly good.

Galena. No mineral collection is complete without a good specimen of galena, the mainstay of the old crystal radio set and one of the leading silver ore minerals of the world. Strangely enough, pure galena contains no silver, being the sulfide of lead, but galena is one mineral in which silver commonly is deposited, either in fractures or by chemical replacement of the lead sulfide itself.

Galena, when untarnished, is a brilliant, silvery-white, metallic-looking mineral showing prominent cleavages arranged in a cubic pattern. Crystals, when present, are generally cubic or octahedral with dull tarnished crystal faces, but few, if any, local occurrences yield crystals. When heated in a blowtorch or before a blowpipe, galena fuses readily, yielding a bead of metallic lead and sulfur dioxide gas. Galena is soft, 2½ on Mohs scale (a little harder than one's fingernail), and very heavy, having a specific gravity of 5.8. The powdered mineral, which is lead gray, easily dissolves in acids.

Lumps of galena have been found along Codornices Creek in Berkeley near the Euclid Avenue culvert. Galena is also found associated with gold and copper minerals at the Palisade mine near Calistoga. Occur-

*Blue glaucophane,
green chlorite
(Jenner)*

*Banded glaucophane–
omphacite–garnet schist
(Jenner)*

Pyrite on quartz

Travertine onyx

Mica schist

Mica–quartz–feldspar gneiss

Hornblende–feldspar pegmatite

Biotite granite

Biotite–hornblende quartz diorite

Gabbro

Banded rhyolite

Andesite

Basalt

Pumice

Scoria

Obsidian

Pillow basalt

Platy flow-basalt

Basalt greenstone with jasper blanket veins (Fort Cronkhite)

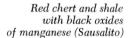

Red chert and shale with black oxides of manganese (Sausalito)

Typical landscape on serpentine rock (Mt. Tamalpais)

Cinnabar

Bornite

Chalcopyrite

Chert and shale interbedded, Claremont formation

Crossbedded pebbly sandstone, Neroly formation

Shale and sandstone interbedded

Gray chert, Franciscan formation

Dolomite crystals (note curved faces)

Magnesite

Calcite marble

Phlogopite mica in pegmatite

Chromite in serpentine matrix

Galena: cubic crystals, cleavage pieces

Asphalt-impregnated sandstone

Mollusc shells in sandstone

*Paleocene boulder conglomerate
on Cretaceous shale*

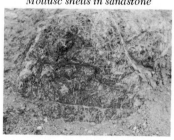

Serpentine rock with picrolite veinlets

Aplite dikes cutting granodiorite

White diatomite in Purisima formation

Azurite,
malachite,
chalcedony

Mariposite–
ankerite rock

Native sulfur

Hematite

rences are occasionally reported from South San Francisco and from the vicinity of Searsville Lake, San Mateo County.

Magnetite and *hematite* are the two most common ore minerals of iron. Although no notable deposits of either mineral are known within the ten Bay counties, both are widely distributed in small amounts in every county. Hematite, the rich, deep-red oxide of iron forms the coloring matter of chert, jasper, and many other rocks and minerals. The heavy black sands found sporadically concentrated on beaches, and watercourses are made up principally of magnetite.

Magnetite is relatively hard (5.5 to 6.5 on Mohs scale), heavy (sp. gr. 5.1), black, and metallic, and is strongly attracted by a hand magnet. The powdered mineral is black as well as the larger pieces. Hematite, although in some cases black, hard, and metallic-looking, always yields a red powder and is feebly attracted by a hand magnet, if at all. Commonly it is soft and earthy; less commonly it is found in glittering black or dark-red scales, like tiny sequins.

Pyrite and *Marcasite*. These are the yellow and white iron pyrites respectively, so familiar to most miners; the yellow variety, common pyrite, is often called fool's gold because of a slight resemblance to the noble metal. The principal use of these pyrites is in the manufacture of sulfuric acid. Both are double sulfides of iron (FeS_2), but marcasite commonly forms at a lower temperature than pyrite. Pyrite generally crystallizes into cubes having striated faces; the pentagonal dodecahedron is also a common crystal form. Marcasite, on the other hand, is either massive or is found in groups of coxcomb-like crystals belonging to the orthorhombic system. The tin-white color and crystal form together are distinctive.

Cubes of pyrite occur with chalcopyrite in the albite-lawsonite schists near Reed's Station, Tiburon Peninsula. Cubes and octahedrons of pyrite are present in the glaucophane schists west of Healdsburg. Both crys-

[41]

talline and massive pyrite are present in the old workings of the Alma mine at Leona Heights, Oakland. Both pyrite and marcasite occur in drusy groups in the dumps of both the Mt. Diablo and St. Johns mercury mines. The latter locality is 2 miles south of Highway 40 and 1 mile north of Sulphur Springs Mountain, or 3½ airline miles northeast of Vallejo.

Chalcopyrite and *bornite*, both sulfides of copper and iron, together form the "peacock ore" of the copper miner. They are two of the most widely distributed and economically important copper ore minerals. Although chalcopyrite is a rich brass yellow and bornite an unusual "horseflesh" red on freshly broken surfaces, both tarnish in air or water vapor with multicolored irridescence much like the colors in a peacock's tail. Bornite, the richer ore of the two, is 63% copper; chalcopyrite is about 37% copper. Because of its greater content of copper, bornite is heavier than chalcopyrite, bornite ranging in specific gravity from 4.9 to 5.4, chalcopyrite from 4.1 to 4.3. Both have about the same hardness on Mohs scale—3 for bornite, 3½ for chalcopyrite. Neither mineral is very often found in good crystals, although partial crystals of chalcopyrite occasionally are found in Bay region localities. These are parts of steep-sided tetragonal pyramids known as sphenoids. Bornite rarely is found in cubic and octohedral crystals. Both chalcopyrite and bornite decompose readily in nitric acid, coloring the acid solution an intense green (copper nitrate). Blue azurite and green malachite are common near surface alteration products of the copper sulfides.

Bornite and chalcopyrite are found in a series of prospect pits in upper Mitchells Canyon, on the northwest slopes of Mt. Diablo A rediscovery of these prospects, originally located in the 1860's, caused some stir in the papers late in 1959, but their economic potential is small. Chalcopyrite is the principal ore mineral in the copper mines near Bolinas, Marin County, from which some ore was shipped during World War I. Partly-formed crystals of chalcopyrite are occasionally found with pyrite, albite, and lawsonite in the glaucophane schists near Reed's Station, Tiburon Peninsula, and in the glaucophane schists west of Healdsburg.

Pyrolusite and *Psilomelane*. These two black oxides of manganese are the principal ore minerals in the many localities scattered throughout the central Coast Ranges. Manganese ore has been produced, chiefly under wartime impetus, from several localities in the San Francisco Bay region. The Ladd-Buckeye district astride the Alameda-Stanislaus County line is the most notable of these. Pyrolusite is the soft, earthy variety most conspicuous in prospect pits and mine workings. Very rarely it is found in tetragonal crystals. Psilomelane is the hard (hardness 5.5 on Mohs scale), compact variety most commonly found in spherical or hemispherical concentric layers—the botryoidal structure of mineralogists. Psilomelane and pyrolusite may occur singly or together, or they may be found as alteration products of and with such minerals as pink rhodochrosite, pink rhodonite, and grayish-yellow bementite. Both minerals decompose in hot hydrochloric acid with evolution of chlorine gas and both are notably heavier than the common, light-colored, rock-forming minerals.

In the San Francisco Bay region manganese minerals are found in notable concentrations only in the cherts of the Franciscan formation. Cherts are well-stratified, thin-bedded red, green, or light-gray chalcedonic rocks laid down on the ocean floor. In addition to the Ladd-Buckeye district, manganese oxides can be collected in road cuts along Highway 1 near Bolinas; in prospect pits along the Arroyo Mocho, Alameda County; on the east slopes of St. John Mountain east of Stonyford, Colusa County; and in the vicinities of Sausalito and the northern end of Golden Gate Bridge.

Chromite is the hard black oxide of chromium and iron, the source-mineral of nearly all commercial chro-

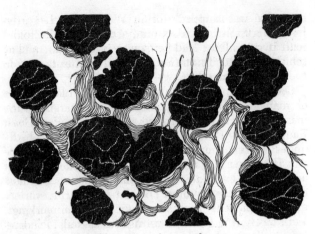

Chromite in leopard ore with serpentine

mium. Bay region mines have supplied chromite during two world wars—as a source of chromium for steel-making and as a tarnish-proof coating for many metals. Chromite is found primarily in peridotite or its serpentinized equivalent, where it may occur in large black masses as spherical kernels in serpentine matrix (leopard ore) or in grains disseminated in serpentine. Octahedral crystals are fairly common. The powdered mineral is dark brown rather than black, and grains are but feebly attracted by a hand magnet—if at all. Cleavage is absent, the hardness is 5.5 on Mohs scale, the specific gravity is 4.1 to 4.9, and the brittle mineral fractures unevenly. As it is heavy and abrasion-resistant, chromite is a common constituent in stream and beach deposits adjacent to serpentine masses.

Leopard ore and massive chromite are present in the old dumps of the Bald Eagle magnesite mine on the west slopes of Red Mountain east of the Arroyo Mocho road 15 miles southeast of Livermore. Most of the serpentine masses in Sonoma and Solano counties contain chromite. The mineral is also found in serpentine near San Geronimo on the north side of Sir Francis Drake

Highway; in the vicinity of Crystal Springs Lakes on Highway 9, San Mateo County; along the Knoxville road 12 miles from Middletown, Napa County; and at the Graves Ranch mine 8 miles northwest of Berryessa Lake, Napa County.

Gold, Silver, and *Platinum.* Small quantities of these native metals have been produced from time to time from the beach sands of the San Francisco Peninsula. They came there largely via the Sacramento–San Joaquin River systems from the Sierra Nevada and Klamath Mountains. Gold and silver have also been produced at the Palisade and Silverado mines near Mount St. Helena, Napa County. There they occur in mineral veins with quartz. In the Palisade mine copper, lead, and iron-bearing minerals are also associated, including pyrite, galena, chalcopyrite, and bornite. Native gold is nearly always alloyed with a little silver; native silver, however, is not necessarily alloyed or associated with gold—many silver mines produce little or no gold. The distinctive color of native gold, together with its great weight, relative softness, and insensibility to acid attack except by aqua regia (a mixture of nitric and hydrochloric acids), make gold one of the most easily identified minerals. Gold crystals generally are cubic or octahedral. Leaves of gold showing a few crystal faces were once found in veinlets in silica-carbonate rock on Keith Avenue in Berkeley. Needless to say, the locality is long since exhausted! Several odd occurrences of gold in San Francisco Bay counties are worthy of note. Specimens of gold deposited on quartz crystals were found in the Manzanita quicksilver mine on Sulphur Creek prior to 1892. Gold, with red garnets and green pyroxene, was found in eclogite on Coyote Creek, Santa Clara County, six miles north of San Martin. Gold nuggets with associated cinnabar have been reported from the Sulphur Springs vicinity in Bear Valley northeast of Borax Lake, Lake County.

Except for the silver found alloyed with gold in placers, most Bay region silver mines yielded silver from either argentiferous galena or argentite. Galena has been described elsewhere. Argentite is dark lead-gray to black, ranges in hardness from 2 to 2½ on Mohs scale, ranges in specific gravity from 7.2 to 7.4, is sectile, and is occasionally found in octahedral crystals or arborescent crystal groups. It has been found in the Palisade mine near Calistoga and in the mines on the southeast slope of Mt. St. Helena.

Platinum, found in California only in placers, is about twice as hard as gold and has about the same weight (specific gravity 14 to 19). Like gold it is soluble only in aqua regia, but is attacked by the arsenical vapors sometimes produced in an assay laboratory, whereas gold is not. Platinum commonly is alloyed with iridium and osmium, but their properties are so similar to those of platinum that they seldom are separated.

MINERALS FROM SEA AND AIR

Minerals such as halite (common salt) and gypsum (hydrous calcium sulfate) form in marshy tide pools along San Francisco Bay by evaporation of sea water. As the annual rate of evaporation in northern Santa Clara County amounts to five or six feet, a flourishing harvest in salt has sprung up. Natural evaporation is fostered in large shallow ponds. As the sea water evaporates, most of the salt crystallizes on the pond floor, the small volume of brine remaining is drained off, and the salt is harvested by specially designed scraper-loader assemblies.

Beautiful crystal groups of halite can be picked up along the Dumbarton Bridge road near Newark during any salt-harvesting period. Similar ponds are maintained along the bay opposite Redwood City. Some of the nicest crystal groups form on twigs and grass stems that have fallen into the ponds. The crystals are either solid or skeletal cubes which cleave readily into cubic fragments. Halite is relatively soft (2½ on Mohs scale),

[46]

Halite (rock salt) crystals: cubic skeletons

tastes salty, and is the most common of the saline minerals.

MINERALS FROM WARM SPRINGS

Many of the warm springs of the San Francisco Bay region—both active and decadent—have deposited minerals of many compositions. The carbonate minerals, aragonite and calcite, commonly accumulate this way as do gypsum and such lesser-known minerals as alunite, epsomite, kalinite, and alunogen. Some of these minerals defy identification except by chemical analysis, but others are readily recognized. Mineral collectors who like to find large numbers of uncommon minerals associated together will do well to visit such places as The Geysers near Cloverdale in Sonoma County and the Sulphur Banks mine on Lower Lake, Lake County.

Crystals, roses, and sheetlike masses of gypsum are found in soil and the underlying shale along Highway 24 near Cowell, just west of the white-sand quarries. It has been an almost inexhaustible source of interesting specimens for several years. Good crystals are also

[47]

found at the Palisade mine near Calistoga. Gypsum is readily recognized by its broad cleavage faces, water-clear transparency, and relative softness (less than one's fingernail). It forms a wide variety of crystal forms belonging to the monoclinic system. Dove-tailed twins are fairly abundant.

MINERALS FROM VOLCANOES

When molten or extremely hot rocks reach the earth's surface from deep within, several things happen which favor mineral deposition. Rapid cooling takes place, high confined pressures are released, and extremely hot vapors and solutions attack the rocks adjacent to the volcanic conduits. All of these events favor accumulation of new minerals in and around the point of emission. As molten material crystallizes, water vapor, carbon dioxide, and a wide variety of other gaseous materials are driven off, so that large numbers of minerals ranging broadly in composition can result. Many of these minerals are similar to those deposited by warm springs; others result from sublimation of superheated materials, such as sulfur and mercury.

San Francisco Bay region residents can observe the effects of waning volcanism at The Geysers, near Cloverdale in Sonoma County, or at Sulphur Bank on Clear Lake, Lake County. Although eruption of molten material has ceased in these vicinities, discharge of hot gaseous material continues from very hot rocks not far beneath the surface. Twenty to thirty minerals can be identified at either locality without excessive effort. Carbonate, sulfate, sulfide, and chloride minerals are likely to predominate. Opal and chalcedony abound and the rarer silica minerals, tridymite and cristobalite, may be detected in small quantities in crystal-lined cavities in volcanic rocks. Sulfur, quicksilver globules, cinnabar, and black stibnite needles are common sublimation products to look for.

ROCKS

Rocks are simply cohesive masses or grains of one mineral or many minerals or, in a few cases such as coal, asphalt, and coquina (cemented seashells), of organic matter. Rocks may form from molten materials by crystallization or solidification to glass, by accumulation of detritus eroded from the earth's surface, by chemical precipitation from water, by chemical interaction between mineral-bearing vapors and solutions and other rocks, or by recrystallization of pre-existing rocks. Igneous rocks are those that have solidified from a molten parent; sedimentary rocks are those accumulated from some transporting medium such as water or air; and metamorphic rocks are those remade from pre-existing rocks.

ROCKS SOLIDIFIED FROM MOLTEN MATERIALS

As rocks come into being from the molten liquid state, several factors influence the resultant end-product. Perhaps the most important of these is time. In general, the longer favorable crystallizing conditions prevail, the more perfectly crystallized and the coarser-grained the rock is likely to be. Water vapor and other gases such as carbon dioxide dissolved in the melt also favor crystallization. Consequently, steady, high-temperature conditions are necessary for good crystallization. Lastly, the temperature-pressure-viscosity relationships must allow free movement of materials within the silicate melt over a long period of time. Molten silicate materials containing little dissolved gas will form a glass, if quickly cooled. One having much dissolved gas will yield a glass froth under rapid cooling unless it is cooled under high confining pressures.

Igneous rocks are conveniently divided into two classes: *plutonic* igneous rocks, typified by granite, are relatively coarse-grained, have formed slowly deep within the earth's crust and consist almost entirely of crystals; *volcanic* igneous rocks have solidified at or close to the earth's surface, are relatively fine-grained (with some exceptions), and the grain size and shape are generally variable within a single hand specimen. Volcanic rocks may consist entirely of glass, partly of glass and partly of crystals, or entirely of crystals. At opposite extremes, among the igneous rocks, are examples that contain no crystals, those made up wholly of very large crystals; and all graduations between the two extremes.

Obsidian is the generally transluscent, relatively light-weight glassy equivalent of granite or rhyolite. *Tachylite* is the black, relatively heavy, glassy equivalent of basalt or gabbro. When molten material contains large amounts of dissolved gases that cannot escape because of high confining pressure, very coarse-grained rock called *pegmatite* is likely to form. Single crystals many tons in weight have been found in some pegmatites, and such rocks are often sources of rare minerals useful to man.

THE GRANITIC ROCKS

With the exception of a little gabbro, found enclosed in or situated at the borders of serpentine masses, and scattered pebbles found in sediments, granitic rocks in the San Francisco Bay region are found west of the San Andreas Fault. Bodega Head, Point Reyes, the Farallones, Montara Mountain, and the core of the Santa Cruz Mountains are the foremost localities. Granitic rocks are named and classified by their coarse-granular, interlocking-crystal texture, by the kinds and properties of the feldspars present, and by the presence of certain characteristic accessory minerals such as hornblende, micas, pyroxenes, and olivine.

Quartz diorite, the principal variety in the Farallones, Point Reyes, and Bodega Head, is made up chiefly of plagioclase (soda-lime) feldspar and quartz. The rock is evenly speckled with dark minerals such as biotite or hornblende, or both. The general aspect of Bay region quartz diorites is light-colored.

Granodiorite, the variety most common on Montara Mountain and the core of the Santa Cruz Mountains, resembles quartz diorite but contains considerable potash feldspar (up to 25% of the rock) as well as plagioclase and quartz. The accessory dark minerals biotite and hornblende are present in amounts simular to most quartz diorites.

Quartz monzonite resembles both granodiorite and quartz diorite but contains approximately equal proportions of plagioclase and potash feldspar. The proportions of quartz (20 to 30%) as well as biotite and hornblende are about the same as for granodiorite and quartz diorite—in the San Francisco Bay region.

True granite, the type in which potash feldspar and quartz are the chief constituent minerals, is rare in the San Francisco Bay region except for the textural varieties aplite and pegmatite. *Granite aplite* is the fine-grained, sugary-textured rock found almost exclusively in dikes cutting granitic or metamorphic rocks. The rock, as a rule, is very light colored. *Granite pegmatite* is the extremely coarse-grained dike equivalent of granite. Muscovite and biotite micas commonly are present; less commonly hornblende and black tourmaline. Both granite aplite and pegmatite are found in dikes cutting the granitic rocks of Point Reyes, Montara Mountain, and the Santa Cruz Mountains.

Gabbro generally stands out in a collection of granitic rocks because of its darker color and greater weight. Black minerals make up over half the volume of many gabbros—amphiboles and pyroxenes being the principal types. Gabbro contains no quartz, and the plagioclase feldspar is high in lime.

Small masses of gabbro commonly are associated with serpentine. The dark minerals in these gabbros generally are diallage or enstatite pyroxene or both. Geologists commonly speak of this type of gabbro as

[51]

norite. A few norites and gabbros contain green olivine and are further designated olivine norite or gabbro, as the case may be. Noritic gabbros crop out along the San Andreas Fault zone near Hayward and in the Fort Ross–Monte Rio district of Sonoma County.

The gabbros that are found associated with granodiorite and quartz diorite commonly contain hornblende or augite instead of diallage or enstatite. Small bodies of hornblende gabbro are found in the Santa Cruz Mountains.

VOLCANIC ROCKS

Volcanic rocks originate from the same molten materials as granitic rocks. Consequently there are volcanic equivalents to match nearly all of the varieties of granitic rocks. Rhyolite has much the same chemical composition as true granite; dacite is the volcanic form of granodiorite; andesite is equivalent to diorite; and basalt corresponds to gabbro. As the volcanic rocks solidified rapidly at or near the earth's surface, only part of the liquid had time to crystallize. Volcanic rocks generally consist of well-defined small crystals of minerals such as feldspars, quartz, amphiboles, pyroxenes, micas, and olivine set in a groundmass of glass or of glass and tiny crystals that are hardly visible to the naked eye.

Certain textures and structures are characteristic of volcanic rocks although they do not always exhibit these. As lava becomes increasingly viscous during cooling, color banding commonly appears parallel to the planes of laminar flow. Furthermore, upon arrival of lava at or near the earth's surface, the confining pressure drops, and bubbles of gas commonly develop in swarms or trains parallel to the flowbands. In unbanded lavas the gas cavities or vesicles, as they are called after solidification, may be disseminated throughout the flow.

If the lava remains permeable to liquids or vapors after solidification, the vesicles may become partially or entirely filled with minerals such as quartz, agate, onyx, opal, zeolites, calcite, chlorite, and siderite. Gas cavities filled with such minerals are termed amygdules. Thus, a lava filled with cavities is called vesicular and one filled with amygdules is termed amygdaloidal.

Basalt and *andesite* are the most abundant volcanic rocks in the San Francisco Bay region. The hills east and west of Napa and those north of Benicia are built predominantly of lava flows of basalt and andesite. Ridge-crests in the Berkeley Hills are apt to be of like character. Ancient partly-altered greenstone basalts are well developed in the North Peak of Mt. Diablo, along the Sir Francis Drake Highway west of San Geronimo, and along Highway 101 north of the Golden Gate Bridge.

Rhyolite forms large, light-colored outcrops in the Berkeley Hills between Little Grizzly Peak and the Mira Vista golf course, in Alum Rock Park east of San Jose, and in the vicinity of Mt. St. Helena. The beautiful flagstones quarried in the Valley of the Moon district of Sonoma County are flow-banded rhyolites in which rare riebeckite amphibole and aegirite pyroxene are commonly found. Mt. St. Helena itself is built up largely of rhyolite flows but is not the site of a former volcanic vent.

SEDIMENTARY ROCKS

Sedimentary rocks are those accumulated by deposition of solids carried in suspension by transporting media such as water or wind. Clay, silt, sand, gravel, cobbles, and boulders may be deposited singly or together to form solid stone. Thus massive clay-stone and platy clay-shale are the ultimate products of clay deposition, and siltstone and sandstone form, from consolidation of silt and sand grains, respectively. Conglomerate is the term given to sedimentary rocks composed of mixtures of pebbles, cobbles, gravel, and sand. Obviously, there will be all gradations between sandstone and boulder conglomerate and gradational varieties are given such names as sandy conglomerate or pebbly sandstone, depending upon which constituents predominate.

Another class of sedimentary rocks are those that form by chemical precipitation from standing bodies of water. Some marine and fresh-water limestones collect in this way as do opaline shales and chert (waterlaid chalcedony). Still other sedimentary rocks form by evaporation of mineral-laden water. The salt and borax deposits found on desert lakebeds and many gypsum deposits form in this way.

ROCKS GATHERED BY RUNNING WATER

One needs only to observe the behavior of streams to realize that they can be responsible for accumulation of vast expanses of fragmental rocks. Although many factors, such as stream gradient, stream volume, and availability of movable material affect the appearance of the detritus deposited, most streamlaid deposits consist of more or less rounded pebbles in uneven, cross-laminated beds. Sandy layers commonly are interbedded with these and sand grains may fill interstices between pebbles and cobbles. In general, the larger the pebbles the more rounded they become when rolled along the streambed. The accompanying sand grains may be quite sharp cornered. In California where hill and mountain topography prevail and rainfall is markedly seasonal and often violent, streamlaid pebble and sand deposits may be punctuated by cobble and boulder beds. In contrast, deposits laid down by large slow-moving rivers crossing plainsland such as the Mississippi are chiefly sand, silt, and clay, with few, if any, cobbles. The gravel beds so extensively excavated at Irvington and Livermore for concrete were deposited by streams that existed hundreds of thousands of years ago. Still older, partially-cemented Miocene stream gravels more than 10 million years old may be seen lapping up onto the still older Cretaceous shales where Highway 50 approaches the Great Valley west of the Byron road junction.

Beach deposits resemble stream deposits except that pebbles are apt to be even more rounded and many of the sand grains may also be rounded and frosted by constant impact with other pebbles and grains in the wave zone. Beach deposits are apt to include molluscan shell fragments, and pebbles commonly show cavities and borings produced by burrowing molluscs.

ROCKS FROM LAKES AND SEAS

Sediments that have accumulated on the floors of lakes and seas have several things in common: they are usually well bedded; the detritus is fairly well sorted within each bed; and, as compared to stream and beach deposits, they generally are fine grained. Clays, silts, and the finer sand sizes predominate although these finer sediments grade toward coarser materials at the margins of the basins of deposition. Beach and stream deposits are aften hard to distinguish between.

In general, marine deposits differ from lake deposits in having a far greater breadth and thickness and in containing remains of salt-water rather than fresh-water organisms. Fresh-water molluscs, for example, are readily distinguished from marine mollusc remains by their much thinner shells.

Chert is a special kind of sedimentary rock conspicuous in the San Francisco Bay region. Most commonly it is of marine origin, but is found in some lake deposits as well. The rock is dense and hornlike and may be almost any color. Red, green, and light-gray hues perhaps are most common. Silica in the form of chalcedony, opal, or jasper is the principal constituent, but the silica may be intermingled with clay, iron oxide, microscopic shells, or even silt. It is believed to form by chemical precipitation from bodies of water into which volcanic mineral waters have been introduced. Microorganisms may play some part in the process of accumulation.

Cherts are generally found thinly and distinctly bedded. Commonly they are interbedded with much thinner layers of shale, but in many instances the chert beds are simply piled one on another. Individual beds commonly are one to two inches thick. Thick-bedded and massive chert bodies are relatively rare in the San Francisco Bay region. In cherts of the Franciscan group of rocks, red iron oxide and black manganese oxide are common constituents of the shale interbeds. In the Corral Hollow district of eastern Alameda County, the manganese-bearing chert shales are extensive enough to be sources of manganese ore.

Inasmuch as well over two-thirds of the land surface of the San Francisco Bay region is underlain by marine deposits, these are to be seen everywhere one travels. The immense thicknesses of Cretaceous shale seen along Altamont Pass and along Highway 40 north of Carquinez Straits, the broad section of Eocene sandstone to be seen along the Kirker Pass and Marsh Creek roads, and the shell-prolific Miocene sandstones accenting the ridges south of Ygnacio Valley road east of Walnut Creek are all typical marine-laid deposits that have become more or less lithified with time. Lithification of sediments involves burial, compaction by loading, and cementation—the latter by redistribution of calcium carbonate or iron oxide in the ground water.

DEPOSITS GATHERED BY THE WIND

The San Francisco Bay region has no very ancient wind-gathered deposits, but there are extensive Recent and Pleistocene dune-sand deposits at many points along our coasts and rivers. Among the Pleistocene wind-laid deposits the buff-colored dune sands seen so prominently along the Antioch shores of the San Joaquin River system are perhaps the most extensive. Several thousand acres of rich agricultural land have been converted from old dunes in the Antioch vicinity. At Pacific Grove in Monterey County dune sands are harvested for use in ceramicware and glass. Hundreds of thousands of tons of dune sand are sold annually to the construction industries by producers scattered along the east shores of Monterey Bay.

Dune sands are fine grained, well sorted as to size, and the grains, if examined with a magnifying glass, are commonly frosted by wind-blast action. Many dunes consist of clean white sand—mainly grains of quartz and feldspar, whereas others such as the Antioch sand and the famous "Sahara" near Yuma, Arizona, contain these minerals plus clay particles and oxides of iron. In some parts of the world dunes are made up entirely of seashell fragments (Daytona Beach, Florida), gypsum (White Sands, New Mexico), or black lava fragments (Hilo, Hawaii). The ancient landscapes 130 million years ago had vast dune-sand areas. The crosslaminated Coconino sandstone near Ashfork, Arizona, and many of the spectacular crossbedded sandstones of Zion National Park are lithified sand dunes.

ROCKS REMADE FROM OTHER ROCKS

These are the most difficult to describe and to understand. As the earth's surface is in almost constant slow motion, with some parts being elevated and others depressed (depression is accentuated by pile-up of sediment in basins or build-up of thick piles of lava), many rocks once situated at the earth's surface become deeply buried within the crust. At these depths, where high temperatures, high pressures, and high shearing stresses commonly prevail, sedimentary, igneous, or even preexisting metamorphic rocks may be completely reformed. Shales become slate or schist, limestone becomes marble, sandstone changes to quartzite, rhyolite is transformed to mica schist, and so on. Under deepseated shearing stresses, platy minerals form and these become concentrated into bands or folia. Fine-grained, foliated metamorphic rocks, composed largely of platy, needlelike or fibrous minerals, are called schists; coarse-grained, banded metamorphic rocks composed of mixtures of the platy as well as the more equidimensional minerals, are called gneiss.

Small patches of schist, gneiss, and marble are found associated with granitic rocks west of the San Andreas Fault from Bodega Head south to San Benito County. The handsomest specimens of gneiss come from the Henry Cowell Home Ranch north of High Street in the city of Santa Cruz. The rock is a striking brown and red biotite-muscovite-quartz-feldspar-garnet gneiss. The garnets range from an eighth of an inch to half an inch in diameter.

THE GLAUCOPHANE SCHISTS

The only other group of metamorphic rocks in the San Francisco Bay region worthy of mention is that vivid and exotic group known as the *glaucophane schists*. These are a multicolored, variably-textured, multimineralic group commonly found at or close to the borders of serpentine intrusions. They probably

Lawsonite crystals in glaucophane schist

have been brought to the surface en masse from great depth by being caught up, along faults, with the more mobile serpentine. In any event they are quite foreign to the rocks with which they are associated. As they can be shown to have originated by transformation of many different rocks, including chert, shale, sandstone, and even greenstone, the process of formation seems to be much more important to formation of the glaucophane schists than the composition of the parent rock.

Although the blue amphibole glaucophane is the characteristic mineral of the glaucophane schists, it is not necessarily found in every specimen one picks up. All sorts of mineral combinations may be present and all sorts of contrasting colors may be evident. As many as 20 minerals can often be identified in a single hand specimen.

Rocks Formed by Living Things

Vast thicknesses of rock within the earth's crust consist entirely or almost entirely of the dead remains of plants and animals. Still other large masses are being formed at the present time by secretions from marine animals such as corals and marine plants such as Lithothamnion algae. Coal is simply the lithified, partly-carbonized remains of plants—leaves, stems, wood, and pollen all contributing to the parent material. Peat is a stage intermediate between plant fiber and coal. Asphalt and petroleum are believed to be natural alteration products from microscopic plant and animal remains.

Many limestones consist almost entirely of seashell fragments (others being chemical precipitates). Still others "grow" on the coral reefs of tropical parts of the world. The famous chalk cliffs of Dover, England, consist principally of the limy protective parts (tests) of countless billions of microscopic animals, called Foraminifera.

Another white, earthy, somewhat chalklike substance

found in great thicknesses is diatomite. Diatomite consists of the siliceous protective coverings of microscopic plants, called diatoms.

Limestone is perhaps the most widely used of the earth's chemical raw materials, being an essential ingredient in manufacture of portland cement, lime, carbon dioxide, steel, glass, and ceramicware, to name a few. Diatomite is mined in enormous quantities in California for insulation, filtration, and refractory products.

Many millions of dollars' worth of coal was mined in the Nortonville-Somersville district north of Mt. Diablo during the period 1860-1905. The coal is of sub-bituminous grade and proved a satisfactory fuel until competition with petroleum products became too great. Peat has been excavated for horticultural purposes in the Bethel Island district of the Sacramento–San Joaquin River delta country for many years.

Shell limestones were formerly quarried for cement along the east side of Highway 29, south of Napa, and in the vicinity of Benicia, but the deposits have been depleted. Unconsolidated beds of oyster shells on the floor of San Francisco Bay near Bay Farm Island have supported one cement plant and several poultry grit operations for several decades. The cement plant at Permanente, near Los Altos, uses a limestone partly formed of the protective coverings of foraminifera

White diatomite and diatomaceous shale crops out north of Highway 40 just west of Pinole. Diatomite is also found extensively along the Los Laurelles grade between the Carmel Valley road and the Monterey-Salinas road down Canyon del Rey. Small deposits of fresh-water diatomite are exposed in roadcuts along the Silverado Trail between Calistoga and St. Helena town where they are found interbedded with other lake deposits and with volcanic rocks.

DISTINGUISHING CHARACTERISTICS

MINERAL	COLOR	LUSTER	STREAK	HARD-NESS	CLEAVAGE	FRACTURE	SPEC. GRAV.	CRYSTALLIZATION	OTHER
1 Actinolite	Grn., blk.-grn.	Glassy	Uncolored	5-6	Good—2 directions	Splintery	3-3.2	Monoclinic; bladed	Often assoc. with pearly talc
2 Aegirite	Grn., red-brn.	Glassy to res.	Pale yel.-gray	6-6.5	Good—2 dir.	Uneven	3.5	Monoclinic; long prisms	In rhyolite with blue riebeckite
3 Agate	Many colors	Greasy, waxy	Uncolored	7	None	Even	2.6	Microscopic	Concentrically banded
4 Agate, Iris	Many pale colors	Greasy, waxy	Uncolored	7	None	Even	2.6	Microscopic	Rainbow irrid. in transm. lt.
5 Agate, Moss	Cloudy with blk. blotches	Greasy, waxy	Uncolored	7	None	Even	2.6	Microscopic	Spots often dendritic
6 Albite	White to gray	Glassy to por-celaneous	Uncolored	6-6.5	Good—3 dir.	Uneven	2.6	Monoclinic; tabular	Cl. surf. have many par. lines
7 Allanite	Red.-brn. to blk.	Pitchy to res.	Lt. brn.	5.5-6	Fair—2 dir.	Uneven, brittle	3-4.2	Monoclinic; tabular to acicular	Commonly radio-active
8 Alunite	White, gry., pnk.	Glassy to porc.	Uncolored	3.5-4	Sometimes good —1 dir.	Uneven to conchoidal	2.5-2.7	Rhombohedral; uncommon	In altered volcanic rocks
9 Alunogen	White, off-white	Porcelaneous	Uncolored	1.5-2	Not conspicuous	Earthy	1.6	Mono. or triclinic; small, fibrous	Soluble in water
10 Amethyst	Purple, violet	Glassy	Uncolored	7	None	Conchoidal to splintery	2.6	Hexagonal prisms and pyramids	In cavities in vol-canic rocks; veins
11 Ankerite	White, gray, tan	Porcelaneous	White	3.5	Good—3 dir.	Uneven, splintery	2.9-3.1	Rhombohedral; uncommon	Mainly in veins
12 Anorthite	White, gray	Glassy to porc.	Uncolored	6-6.5	Good—3 dir.	Uneven	2.75	Monoclinic; tabular to prismatic	Cleavages show fine, par. lines
13 Anthophyllite	Gray, brn.-gray	Glassy to dull	White	5.5-6	Good—2 dir.	Splintery	2.8-3.2	Orthorhombic-bladed	Commonly assoc. with serpentine
14 Antigorite	Green, blk.-grn.	Greasy to res.	White	2.5-4	Inconspicuous	Uneven; splintery	2.5-2.6	Monoclinic; uncommon	Joint surfaces often polished
15 Aragonite	White, cream, pale colors	Glassy to res.	Uncolored	3.5-4	Good—2 dir.	Uneven	2.9	Orthorhombic; often columnar	Common in veins and cavity lngs.
16 Argentite	Lead-gray to blk.	Metallic	Blk. to lead-gry.	2-2.5	Inconspicuous	Subconchoidal	7.2-7.3	Cubic or octahedral	Surface often coated with blk. oxide
17 Augite	Blk., brn.-blk.	Glassy to dull	Gray,gray-grn.	5-6	Good—2 dir. at right angles	Uneven	3.2-3.6	Monoclinic; stubby prisms	In granitic rks. and lavas

MINERAL	COLOR	LUSTER	STREAK	HARD-NESS	CLEAVAGE	FRACTURE	SPEC. GRAV.	CRYSTALLIZATION	OTHER
18 Azurite	Azure blue, dk. bl.	Vitreous to adamantine	Blue	3.5-4	Fair—2 dir.	Conchoidal	3.8-3.9	Monoclinic; tabular	Usually with green malachite
19 Bastite	Green	Glassy	Light green	3.5-4	Good—1 dir.	Splintery	2.5-2.6	Monoclinic; tabular	Distinct xls. in serpentine
20 Bementite	Pale yel. to gry.	Resinous	Uncolored	2.9-3	Inconspicuous	Even to conch.	4-6	Orthorhombic	With rhodonite & blk. oxides
21 Biotite	Blk. to dk. brn.	Glassy to pearly	Uncolored	2.5-3	Perfect—1 dir.	Uneven, scaly	2.7-3.1	Monoclinic-pseudo-hexagonal	Splits into flex.-plates
22 Bloodstone	Red spots in blk. matrix	Waxy		7	None	Even, conch.	2.6	None	Matrix green to nearly blk.
23 Bornite	Dk. red to purple peacock tarnish	Submetallic to metallic	Gray to black	3	Inconspicuous	Uneven to conch.	4.9-5.4	Cubic-rare	Commonly has peacock irid.
24 Bowlingite	Amber, yel.-or.	Resinous	Pale yellow	1-2	None	Uneven to earthy	2.2-2.3	None	Yellow spots in altered basalt
25 Calcite	Any color but usually white	Glassy to res.	White	3	Perfect—3 dir.	Conch. to splintery	2.7	Hexagonal rhombohedral	Cl. often shows fine par. lines
26 Carnelian	Orange-red	Waxy	Uncolored	7	None	Even to conch.	2.6	None	Translucent to transparent
27 Chalcedony	Bluish-gray to white	Greasy, waxy	Uncolored	7	None	Even to conch.	2.6	None	Translucent
28 Chalcopyrite	Yellow with peacock irridescence	Metallic	Grn.-blk.	3.5	Poor—1 dir.	Uneven	4.1-4.3	Sphenoidal, octahedral	Softer than iron pyrite and darker
29 Chlorite	Grn., blk.-grn.	Glassy to pearly	Unc. to pale grn.	2-2.5	Perfect—1 dir.	Uneven, scaly	2.6-2.8	Pseudohexagonal	Micaceous
30 Chromite	Black	Metallic	Dk. brn.	5.5	None	Uneven, hackly	4.1-4.9	Octahedral	Associated with serpentine
31 Chrysoprase	Apple green	Greasy, waxy	Uncolored	7	None	Even to conch.	2.6	None	In veins in serpentine
32 Chrysotile	White to pale grn.	Pearly to glassy	Uncolored	3-4	Not apparent	Breaks into fibers	2.5-2.6	Fibrous	Fibers flexible
33 Cinnabar	Red, dark red	Adamantine to submetallic	Red, purplish	2-2.5	Poor—1 dir.	Uneven	8-8.2	Rhombic, tabular	Massive to earthy
34 Citrine	Yellow	Glassy	Uncolored	7	None	Conchoidal	2.6	Hex.-prisms and Pyramids	In veins or xl.-lined cavities

#	Name	Color	Luster	Streak	H	Cleavage	Fracture	G	Crystal form	Remarks
35	Clinozoisite	Gray, greenish	Glassy to greasy	Uncolored	6–7	Good—2 dir.	Uneven	3.2–3.5	Monoclinic-long prisms	In metamorphic rocks only
36	Cristobalite	White	Glassy	Uncolored	7	None	Uneven	2.2–2.3	Octahedral	In small cavities in lavas
37	Crossite	Blue, blue-blk.	Glassy to pearly	Grayish-blue	6–6.5	Good—2 dir.	Uneven	3.1–3.2	Monoclinic-acicular	Identifiable only by optical prop.
38	Diallage	Dk. grn., brn.	Bronzy to dull	Gray-green	4	Good—1 dir.	Uneven to conch.	3.2–3.3	Monoclinic-stubby prisms	Pseudofoliated
39	Diopside	Grn., gray-grn.	Glassy	White, gray	5–6	Good—2 dir.	Uneven to conch.	3.2–3.4	Mono.-prismatic	Common in crystalline limestone
40	Dolomite	White, pale colors	Glassy, pearly, porcelaneous	White	3.5–4	Good—3 dir.	Subconchoidal to uneven	2.8–2.9	Rhombohedral	Crystal faces commonly curved
41	Enstatite	Grn., gray-grn.	Pearly, vitreous	Unc., grayish	5.5	Good—1 lir.	Uneven, brittle	3.1–3.3	Orthorhombic	Commonly altered to serpentine
42	Epidote	Pistachio-grn.	Glassy	Unc., grayish	6–7	Good—2 dir.	Uneven, brittle	3.2–3.5	Monoclinic-long prisms	Usually found in good crystals
43	Epsomite	White	Glassy to dull	White	2–2 5	Good—1 dir.	Fine-granular to earthy	1.7	Fibrous crusts	Bitter, salty taste
44	Fayalite	Yel., brn., blk.	Glassy	Pale-yel., brn.	6.5	Good—2 dir.	Uneven	4.1	Monoclinic-small prisms	In cavities in lava
45	Flint	Gray, brown	Dull to waxy	Uncolored	7	None	Conchoidal	2.6	None	Usually associated with chalk
46	Forsterite	Green, yellow	Glassy	Uncolored	6–7	Poor to absent	Uneven	3.2–3.3	Orthorhombic-us.-rounded grains	In xline limestones and dolomites
47	Galena	Lead-gray	Metallic	Lead-gray	2.5–2.7	Cubic-conspic.	Even	7.4–7.6	Cubic, octahedral	Commonly assoc. with pyr. & sphal.
48	Garnet, Almandite	Red, brown	Glassy	Unc., pale-col.	6.5–7.5	None	Uneven to conch.	4.2	Dodecahedral, trapezohedral	Commonly in schists
49	Garnet, Spessartite	Red, lavender	Glassy	Uncolored	6.5–7.5	None	Uneven to conch.	4.2	Dodecahedral, trapezohedral	In metamorphic rocks
50	Gastaldite	Blue, blue-blk.	Glassy	Grayish, bluish	6–6.5	Good—2 dir.	Uneven to conch.	3.1–3.2	Monoclinic-long prisms	Identifiable only by optical prop.
51	Glaucophane	Blue, blue-blk.	Glassy	Grayish, bluish	6–6.5	Good—2 dir.	Uneven to conch.	3.1–3.2	Monoclinic-long prisms	Common in Coast Range schists
52	Gold	Deep yellow	Metallic	Golden	2.5–3	None	Hackly	19.3	Cubic, octahed.	Rarely in xls.
53	Gypsum	Uncolored, white or pale colored	Glassy, pearly	White	1.5–2	Perfect—2 dir.	Uneven, conch.	2.3	Monoclinic, tab.	Dove-tailed twins common

DISTINGUISHING CHARACTERISTICS

MINERAL	COLOR	LUSTER	STREAK	HARD-NESS	CLEAVAGE	FRACTURE	SPEC. GRAV.	CRYSTALLIZATION	OTHER
54 Halite	Uncolored, white	Glassy, pearly	Uncolored	2.5	Good — cubic	Conch., uneven	2.1–2.6	Cubic	Salty taste
55 Hematite	Dk. red to black	Metallic	Red, red-brn.	5.5–6.5	None	Uneven, earthy	4.9–5.3	Hex., often earthy	Usually massive and impure
56 Hornblende	Blk, dk. brn.	Glassy	Gray-green	5–6	Good — 2 dir.	Uneven to conch.	2.9–3.4	Monoclinic-long prisms	Sometimes acicular
57 Iddingsite	Red-brown	Resinous, waxy	Red-brown	3	Not conspicuous	Conch., crumbly	2.5–2.8	Orthorhombic	In basalt
58 Jadeite	Green to white	Greasy, waxy; xls. glassy	Uncolored	6–7	Good — 2 dir.; us. inconspicuous	Splintery	3.3–3.5	Monoclinic	Us. in tough, inter-woven fibers
59 Jasper	Red, yel., grn.	Glassy to dull	Pale to un-colored	7	None	Conch. to uneven	7	None	Takes high polish
60 Jasper, orbicular	Orbs of one color in other matrix	Glassy to dull	Pale colors	7	None	Conch. to uneven	7	None	Takes high polish
61 Kalinite	White, off-white	Glassy to pearly	White, unc.	2	Not apparent	Earthy	1.7	Mono-fibrous	Bitter, acrid taste
62 Labradorite	White, gray	Glassy to pearly	White	5–6	Good — 2 dir.	Uneven to conch.	2.6–2.7	Triclinic-tabular	Cleavages show fine, par. lines
63 Lawsonite	White, blue-gray	Glassy to greasy	White	7–8	Good — 2 dir.	Uneven to conch.	3	Orthorh.-tabular	In glauc. schist
64 Magnesite	White to lt. brown	Porcelaneous	White	3.5–4.5	Usually absent	Conchoidal	3–3.1	Rhombic but us. not apparent	Commonly botry-oidal
65 Magnetite	Black	Metallic	Black	5.5–6.5	None	Uneven to hackly	5.1	Isometric-us. octahedral	Strongly magnetic
66 Malachite	Deep green	Glassy to silky	Light-green	3.5–4	1-dir. — not consp.	Uneven, brittle	3.9–4	Monoclinic	Commonly botry-oidal or banded
67 Marcasite	Pale brass-yel.	Metallic	Brownish blk.	6–6.5	None	Hackly	4.8–4.9	Orthorhombic-tabular coxcombs	Often assoc. with opal and chalced.
68 Mariposite	Apple-green	Glassy, greasy	White	2.5–3	Perfect — 1 dir.	Uneven	2.7–2.8	Monoclinic-pseudo-hexagonal	Thin, flex. plates
69 Metacinnabarite	Black	Metallic	Black	3	None	Uneven, brittle	7.7	Isometric but usually massive	Sectile — commonly with cinn.
70 Microline	Pink, white salmon	Glassy	White	6–6.5	Perfect — 3 dir.	Uneven	2.5	Triclinic-stubby prisms	Shows many thin, wavy lines

No.	Name	Color	Luster	Streak	Hardness	Cleavage	Fracture	Crystal System	S.G.	Remarks
71	Muscovite	Unc. to pale grn.	Glassy	Uncolored	2–2.5	Perfect – 1 dir.	Uneven	Monoclinic-pseudo-hexagonal	2.7–3	Splits into thin, flexible plates
72	Olivine, common	Green to brown	Glassy	Uncolored	6.5–7	Not conspic.	Uneven	Orthorhombic	3.2–3.3	Xls. us. in cavities in basalt
73	Omphacite	Grass-green	Glassy	Uncolored	6–6.5	Good – 2 dir.	Uneven, splintery	Mono.-stubby prisms	3.5	With red garnet
74	Onyx	Many colors	Greasy, porcelaneous	White, unc.	7	None	Conchoidal	None	2.6	Banded-flat, parallel bands
75	Opal	Many colors	Fiery to waxy	White, unc.	5.5–6.5	None	Conchoidal	None	2.1–2.2	Transparent to translucent
76	Orthoclase	Pink, white, salmon	Glassy	White, col.	6–6.5	Perfect – 3 dir.	Uneven, splintery	Monoclinic	2.5	Common in granitic rocks
77	Peridot	Green	Glassy	Uncolored	6.5–7	Not conspicuous	Uneven	Orthorhombic	3.2–3.3	Stubby prisms and pyramids
78	Phlogopite	Pale yel., bronzy	Glassy	Uncolored	2.5–3	Perfect – 1 dir.	Uneven	Monoclinic-pseudo-hexagonal	2.7–2.8	In thin, flex. plates
79	Picrolite	Green to white	Greasy, waxy	Uncolored	4	Fair – 1 dir.	Uneven, splintery	Mono.-fibrous	2.5–2.6	Brittle fibers
80	Piedmontite	Red, pink	Glassy	White, pale-pink	6.5	Good – 2 dir.	Uneven, brittle	Monoclinic-long prisms	3.4	In veins, lavas
81	Platinum	Silver-white	Metallic	Silvery	4–4.5	None	Hackly	Isometric	14–19	Grains and nuggets in placers
82	Psilomelane	Black	Metallic to dull	Black	5–7	None	Conch., uneven	Not conspicuous	3.3–4.7	Massive to botryoidal
83	Pumpellyite	Greenish, yel.	Glassy to greasy	Uncolored	5.5	Fair – 1 dir.	Uneven, splintery	Orthorhombic, platy to fibrous	3.2	In metamorphic rocks
84	Pyrite	Brass-yellow	Metallic	Greenish blk.	6–6.5	None	Conch. to hackly	Isometric-cubes or pyritohedrons	5	Assoc. with gold, galena, sphal.
85	Pyrolusite	Black	Earthy, dull	Black	2–2.5	None	Uneven to earthy	Tetragonal, rare	4.7–4.8	Common black coating on rocks
86	Quartz	Uncolored, white	Glassy	Uncolored	7	None	Conchoidal	Hexagonal	2.6	Forms good xls.; also massive
87	Quartz, rose	Pink	Glassy	Uncolored	7	None	Conchoidal	Hexagonal	2.6	Us. massive
88	Quartz, smoky	Smoky, gray	Glassy	Uncolored	7	None	Conchoidal	Hexagonal	2.6	Xls. or massive
89	Quicksilver	Tin-white	Metallic	None	Liquid	None	None	Liquid	13.6	Small globules

DISTINGUISHING CHARACTERISTICS

	MINERAL	COLOR	LUSTER	STREAK	HARD-NESS	CLEAVAGE	FRACTURE	SPEC. GRAV.	CRYSTALLIZATION	OTHER
90	Rhodochrosite	Pink, yel., gray	Glassy	White	3.5–4.5	Perfect – rhombic	Uneven, brittle	3.4–3.6	Hexagonal-rhombo-hedral	Assoc. with blk. oxides of Mn.
91	Rhodonite	Pink	Glassy	White	5.5–6.5	Good – 2 dir.	Uneven, brittle	3.4–3.7	Triclinic	Massive-gran. to columnar-fibrous
92	Riebeckite	Dark-blue	Glassy	Blue-gray	5–6	Good – 2 dir.	Uneven	3.4	Monoclinic	Slim prisms
93	Siderite	Brn., yel.-brn.	Glassy to dull	White, pale-yel.	3.5–4	Good – 3 dir.	Uneven	3.8	Orthorhombic	Usually massive
94	Silver	Silver	Metallic	Silver-white	2.5–3	None	Hackly	10.5	Isometric-rare	Malleable, ductile
95	Soapstone	White to black	Porcelaneous to dull	White	1–1.5	Not apparent	Smooth	2.7–2.8	Massive	Sectile, soapy feel
96	Sphene	Yel., brn., pink	Glassy	Uncolored	5–5.5	Good – 1 dir.	Uneven, brittle	5–5.5	Monoclinic	Wedge-shaped xls.
97	Steatite	White, gray, grn.	Waxy	White	1–1.5	Not apparent	Even, smooth	2.7–2.8	Massive	Soapy feel
98	Stibnite	Lead-gray	Metallic	Lead-gray	2	Good – 1 dir.	Uneven, platy	4.5–4.6	Orthorhombic-long prisms	Crystal faces striated
99	Stilpnomelane	Brn., blk., yel.	Submetallic to glassy	Off-white	3–4	Perfect – 1 dir.	Uneven	2.7–2.7	Monoclinic	Flexible plates
100	Sulfur	Yellow, brown, blk.	Resinous	White, off-white	1.5–2.5	Fair – 3 dir.	Conch. to uneven	2.0–2.1	Orthorhombic	Massive, brittle
101	Talc	White, gray, grn.	Waxy, pearly	White	1–1.5	Good – 1 dir.	Even, smooth	2.7–2.8	Monoclinic-platy	Soapy feel; sectile
102	Tremolite	White, cream, gray	Glassy to silky	White	5–6	Good – 2 dir.	Splintery	2.9–3.2	Monoclinic-bladed	Often needle-like or asbestiform
103	Tridymite	Colorless	Glassy	Uncolored	7	None	Uneven, splintery	2.3	Orthorhombic-platy	Cavities in lava
104	Zeolite group	Usually white	Glassy, silky	White	3.5–5.5	Usually good	Usually splintery	2.1–2.5	Many types	Us. in volc. rks.
105	Zoisite	Gray, white	Glassy	White	6–6.5	Good – 1 dir.	Uneven to splintery	3.2–3.4	Orthorhombic	With glaucophane in schist

ACTIVITIES

WHERE TO GO

The San Francisco Bay region is richly endowed with geologic and mineralogical features nearly everywhere one goes. Roadcuts along the main routes of travel are among the best places to observe unweathered rocks and minerals. Beds of trunk streams and shingle beds on pebbly beaches are particularly good places to look for rocks and minerals that will take a polish. A tremendous variety of rocks and minerals may be picked up in the vicinity of Mt. Diablo, on Tiburon Peninsula north of Reed's Station, in the Richmond Hills along Arlington Avenue near the Cutting Boulevard intersection, along the Arroyo Mocho in the vicinity of Red Mountain, in the glaucophane schist country west of Healdsburg, and in the vicinity of Geyserville in Sonoma County.

The State Division of Mines in the Ferry Building, San Francisco, has published two books that do much toward answering the question of where to go for minerals and rocks. Bulletin 136, *Minerals of California*, lists most of the known Bay region minerals in a treatise arranged alphabetically by mineral species. Bulletin 154, *Geologic Guidebook to the San Francisco Bay Counties*, has detailed road logs for all of the main routes radiating from San Francisco and Oakland, as well as for many minor routes of travel. It also covers the various facets of Bay region geology and the adjacent mineral deposits and mineral industries. Other guidebooks of more technical nature may be found in the appended reference list.

WHAT TO TAKE

To make the most of a jaunt to look at and collect rocks and minerals the following list of equipment is helpful:

[67]

1. Hand magnifying glass—10-to-20-power.
2. Geologist's pick or bricklayer's hammer.
3. Paper or cloth collecting bags (paper bags are nice to write on but tear easily).
4. Knapsack and tissue paper for packing specimens.
5. Roll of adhesive tape for numbering specimens.
6. Notebook and pencil or pen.
7. Selected reference books.
8. U. S. Geological Survey topographic maps (obtainable at a nominal cost from the U.S.G.S. at 630 Sansome Street, San Francisco). These maps are of great value in locating mineral localities and determining the best approach routes.

To this list might be added an acid bottle, hand magnet, hardness set, streak plate (unglazed porcelain), and pocket knife—for those who like to make their identifications in the field. Binoculars and photographic equipment are often desirable, and don't forget your light meter.

What to Bring Back

Try to limit the size of the specimens you collect, or storage and display problems will quickly become acute. Unless you covet large cabinet-type specimens, pieces of 3x2x1-inch dimensions generally are large enough. Some favor 1-inch cubes (approximately). Select representative, unweathered material obtainable with a little extra work. Where possible, get well-crystallized material—more can be learned from such specimens and they usually are the most attractive. Take adequate notes at the mineral locality—you might want to get there again or describe it to a friend. Label all specimens; an unlabelled, unidentified collection is almost worthless. Specimen labels should ultimately include the name of the mineral or minerals, rock or rocks, the chemical composition(s), the locality collected (preferably by section, township, and range), the collector's name, and the date of the visit.

Remember that minerals and rocks, unlike plants and animals, are not easily regenerated. With the burgeoning population and the great increase in the number of rock and mineral collectors, Bay region localities will be quickly exhausted unless some conservation is practised. Collect only what you need and leave some for someone else.

REFERENCES

English, G. L., and Jensen, D. E., *Getting Acquainted with Minerals*, 117 pp., Hart Book Company, New York, 1958.

Fenton, C. L., and Fenton, M. A., *The Rock Book*, 348 pp., Doubleday, Doran and Co., Inc., Garden City, New York, 1942.

Fritzen, D. K., *The Rock Hunter's Field Manual*, 207 pp., Harper and Brothers, New York, 1959.

Hurlbut, C. S., *Dana's Manual of Mineralogy*, 17th ed., 609 pp., John Wiley and Sons, Inc., New York, 1959.

Jenkins, Olaf P., *et al.*, *Geologic Guidebook of the San Francisco Bay Counties: California Division of Mines and Geology Bulletin 154*, 386 pp., 1951.

Jensen, D. E., *My Hobby Is Collecting Rocks and Minerals*, 117 pp., Hart Book Company, New York, 1955.

Murdoch, Joseph, and Webb, R. W., *Minerals of California: California Division of Mines and Geology Bulletin 173*, 452 pp., 1956 Supplement, 64 pp., 1960.

Pearl, R. M., *Mineral Collector's Handbook*, 297 pp., Mineral Book Co., Colorado Springs, Colorado, 1948.

Pearl R. M., *Rocks and Rock Minerals*, 260 pp., Barnes and Noble, New York, 1956.

Pirrson, L. V., and Knopf, Adolph, *Rocks and Rock Minerals*, John Wiley and Sons, Inc., New York, 1947.

Rogers, A. F., *Introduction to the Study of Minerals*, 601 pp., McGraw-Hill Book Co., New York, 1937.

Woolf, D. O., *The Identification of Rock Types*, 11 pp; U. S. Dept. of Commerce, Bureau of Public Roads, U. S. Gov't Printing Office, Washington, D. C., 1950.

GLOSSARY

Acicular—needlelike

Adamantine—diamondlike luster; reflected light is broken up into the colors of the spectrum

Botryoidal—having a surface consisting of small, rounded prominences

Conchoidal—having a surface consisting of curved, concave depressions, as in broken glass or rosin

Coxcomb—serrated like the single comb in fowls

Dodecahedron—a symmetrical twelve-sided figure each face of which is a rhomb, and having three equal axes of symmetry at right angles to one another

Ductile—capable of being drawn into wire

Foliation—platy cleavage in rocks and minerals; the plates generally other than flat

Geode—a large cavity in rock lined with crystals of one or more minerals

Hackly—a surface having sharp, jagged projections

Hexagonal—a system of crystals or solid figures having three axes of equal length and angular spacing and a fourth axis at right angles to the plane of the other three; cross sections usually six-sided

Isometric—a system of crystals or solid figures having three axes of equal length lying at right angles to one another

Malleable—capable of being hammered out into very thin sheets

Monoclinic—a system of crystals or solid figures having three axes of unequal length one of which is inclined to the plane containing the other two

Octahedron—an eight-sided crystal or solid figure consisting of two pyramids joined base to base

Orthorhombic—a system of crystals or solid figures having three axes of equal length all disposed at right angles to one another

Prism—in crystallography a form whose bases are similar, equal, and parallel polygons and whose lateral faces are parallelograms

Pyritohedron—or pentagonal dodocahedron—a symmetrical crystal having twelve faces each of which is a pentagon

Resinous—the color and texture of amber or rosin

Rhombic—a diamond-shaped figure, i.e., an equilateral, four-sided figure having oblique angles.

Rhombohedral—a crystal or solid figure bounded by six like faces, each a rhomb

Sectile—capable of being cut into pieces with a knife without falling to pieces or being reduced to powder

Sphenoid—a four-faced solid each face of which is an isosceles triangle

Subchonchoidal—having fracture surfaces on which shell-like depressions are imperfectly or very sparsely developed

Tenacity—the ability of a mineral to resist disruption; almost synonymous with elasticity

Tetragonal—a system of crystals or solid figures having three axes of symmetry mutually at right angles to one another; two of these of equal length; the third longer or shorter than the other two

Tetrahedron—a crystal or solid figure having four sides each of which is an equilateral triangle

Trapezohedron—a crystal or solid figure having twenty-four equally developed faces, each a trapezoid

Triclinic—a system of crystals or solid figures having three unequal axes all mutually oblique to one another